11+ Practice Papers

Mixed
Multiple-Choice
Pack Four 11A

Read these instructions before you start:

- There are three sections in this paper.
- You have **50 minutes** in total to complete the whole paper. The time allowed for each section is given at the beginning of that section.
- There are **80 questions** in this paper and each question is worth **one mark**.
- You may work out the answers in rough on a separate sheet of paper.
- Answers should be marked on the answer sheet provided, not on the practice paper.
- Mark your answer in the column marked with the same number as the question by drawing a firm line clearly through the box next to your answer.
- If you make a mistake, rub it out as completely as you can and mark you new answer. You should only mark one answer for each question.
- Work as quickly and carefully as you can.
- If you find a question difficult, do **NOT** spend too much time on it but move on to the next one.
- **Calculators and protractors are not allowed.**

Moon Tuition
making the most of your potential
www.moontuition.co.uk

Part I

Numerical Reasoning - 25 minutes

Question 1 - 5

London	Beijing	New York	Paris	Sydney
12:00	20:00	07:00	13:00	23:00

The table above shows the time difference among 5 cities around the world.

The table below is the answer bank for question 1 to 5.
Choose your answers from the answer bank and mark your choice in the answer sheet.

A	B	C	D	E	F	G	H
1 hour	8 hours	3 hours	6 hours	11 hours	05:00	12:00	23:00

1 What is the time difference between Beijing and London?

---

2 What is the time difference between Paris and London?

---

3 What is the time difference between New York and Paris?

---

4 What is the time difference between Sydney and Beijing?

---

5 Lucas is going to travel from London to Beijing. If he leaves London at 09:00, the flight journey from London to Beijing is 12 hours, what local time will it be when he arrives at Beijing?

Question 6 - 12

orange	apple	banana	kiwi	watermelon
48p	99p	45p	35p	£2

The table above shows the cost of buying each of the five different fruits.

The table below is the answer bank for question 6 to 12.
Choose your answers from the answer bank and mark your choice in the answer sheet.

A	B	C	D	E	F	G	H	I	J	K	L
3	10	5	20	£1.60	£1.80	15	£1.65	£5.98	£1.40	£3.60	£4.60

6 How many apples can you buy with a £5 note?

7 How many oranges can you buy with a £10 note?

8 How many bananas can you buy if you spend all £3.35 on some bananas and watermelons?

9 How much more expensive to buy a watermelon than a kiwi?

10 How much does it cost to buy 2 apples and 2 watermelons?

11 The watermelon's price is reduced by 30 % in a sale, how much does it cost to buy a watermelon in the sale?

12 How much change will you get if you buy 4 kiwis with a £5 note?

Question 13 - 19

Alex	Tom	Richard	Angus	Ian	Dachi
1.83m	1.75m	1.77m	1.78m	1.63m	1.63m

The table above shows the heights of six children.

The table below is the answer bank for question 13 to 19.
Choose your answers from the answer bank and mark your choice in the answer sheet.

A	B	C	D	E	F	G	H	I	J
161cm	5cm	20cm	50cm	15cm	1.63m	1.78m	1.76m	1.88m	1.80m

13 How much taller is Alex than Angus?

14 What is the range of the heights?

15 What is the mode of the heights?

16 What is the median of the heights?

17 Dachi has grown 2cm since last year. How tall was he last year?

18 If Tom grows another 50mm by his next birthday, how tall will he be on his next birthday ?

19 How much shorter is Ian than Angus?

Question 20 - 26

Milk	Flour	Eggs	Water
200ml	480g	2	1.6 litres

The table above shows the receipt to make a cake for 4 people.

The table below is the answer bank for question 20 to 26.
Choose your answers from the answer bank and mark your choice in the answer sheet.

A	B	C	D	E	F	G	H	I	J
400ml	300ml	4	500ml	3	720g	3.2 litres	2.4 litres	0.96kg	0.6 litres

20 How many eggs are needed to make a cake for 6 people?

21 How much milk is needed to make a cake for 6 people?

22 How much water is needed to make a cake for 8 people?

23 How much water is needed to make a cake for 6 people?

24 How much flour is needed to make a cake for 8 people?

25 How much flour is needed to make a cake for 6 people?

26 How much milk is needed to make a cake for 12 people?

Question 27 - 30

27 What is the next number in the sequence below?

2 7 9 16 25

A) 38 B) 42 C) 41 D) 39 E) 45

---

28 In a Tennis club, there are 4 boys for every 2 girls. There are 24 children altogether. How many girls are there?

A) 4 B) 8 C) 2 D) 10 E) 18

---

29 What is the missing number in this table?

IN	3	6	7	9	11
OUT	5	8	9		13

A) 17 B) 12 C) 10 D) 11 E) 21

---

30 A box can hold 8 eggs. How many boxes are needed for 100 eggs?

A) 7 B) 11 C) 12 D) 14 E) 13

THE END OF NUMERICAL REASONING PART

Part II

Verbal Reasoning - 15 minutes

Section 1 - Cloze Test

Complete the sentence by selecting one word from the options A-E. Mark your answer on the answer sheet.

(**31** - A: Standing, B: Stands, C: Stood, D: standing , E: Been stood) on the bank, I felt sick at my stomach. I never had liked (**32** - A: a high, B: tall, C: heights, D: hights , E: high). I heard Mr (**33** - A: pentland, B: Pentland's, C: Pentland, D: pentland's , E: Pentlands) voice above the roar of the water. "Stomp your feet now. Get 'em warm.Then come on - but first scrape your boots, then hoist your skirts."
Mechanically I did as (**34** - A: they, B: she, C: when, D: he , E: him) was directing me, then took a deep (**35** - A: breath, B: water, C: pain, D: sight , E: look) and put one foot on the log. It swayed a little and my boot sent a piece of bark flying into the water. I (**36** - A: take, B: takes, C: took, D: taken , E: has taken)a few steps, shut my eyes, then (**37** - A: look, B: opened, C: open, D: looked , E: thought) them again. Another step. Perhaps if I kept on looking at Mr Pentland waiting for me on the other bank – step - or kept my eyes on my valise - step - and did not once look below me - step - the sound of water became a roar in my ears. That meant I (**38** - A: must, B: did, C: soon, D: definitely , E: slowly) be about halfway now.

PLEASE CONTINUE TO THE NEXT PAGE

Section 2 - Synonyms

Select the word which has the SIMILAR meaning as the word in bold. Mark your answer on the answer sheet by choosing one of the options A-E.

39

vanquish

A. quiet B. defeat C. sound D. deduct E. make

40

reveal

A. hide B. away C. weak D. show E. consider

41

retain

A. weave B. courage C. hold D. listen E. carry

42

vigilant

A. thoughtful B. iconic C. violent D. discourage E. watchful

43

flawless

A. clever B. perfect C. fault D. heavy E. feeble

44

secluded

A. concerned B. fixed C. move D. unhappy E. isolated

PLEASE CONTINUE TO THE NEXT PAGE

45

vague

A. rough B. unclear C. meagre D. distant E. empty

46

deceive

A. trick B. sore C. dash D. touch E. simplify

47

capitulate

A. charge B. force C. surrender D. catch E. escape

48

diligent

A. clear B. smooth C. far D. touching E. industrious

49

vivid

A. long B. bright C. picky D. affluent E. taken

50

coarse

A. rough B. length C.worry D. tighten E. mixed

PLEASE CONTINUE TO THE NEXT PAGE

Section 3 - Shuffled Sentence

The following are shuffled sentences. Select the word that needs to be removed so that you can rearrange the rest of the words to make a sentence. Mark your answer on the answer sheet by choosing one of the options A-E.

51 of eight-year-olds half children decay tooth for have

A) decay B) children C) of D) for E) have

---

52 a disease decay tooth take is serious

A) take B) tooth C) a D) serious E) is

---

53 signs improvement of there but happen are

A) signs B) of C) there D) but E) happen

---

54 10000 nearly children must surveyed were

A) nearly B) must C) surveyed D) were E) children

---

55 very the overall we figures encouraging were

A) we B) figures C) encouraging D) were E) overall

---

56 a reduce the amount of snacks sugary drinks and children's diet in

A) reduce B) and C) diet D) a E) amount

THE END OF VERBAL REASONING PART

Part III

Non Verbal Reasoning - 10 minutes

Section 1

Which one of the five figures is most unlike the other four? Mark its letter on the answer sheet.

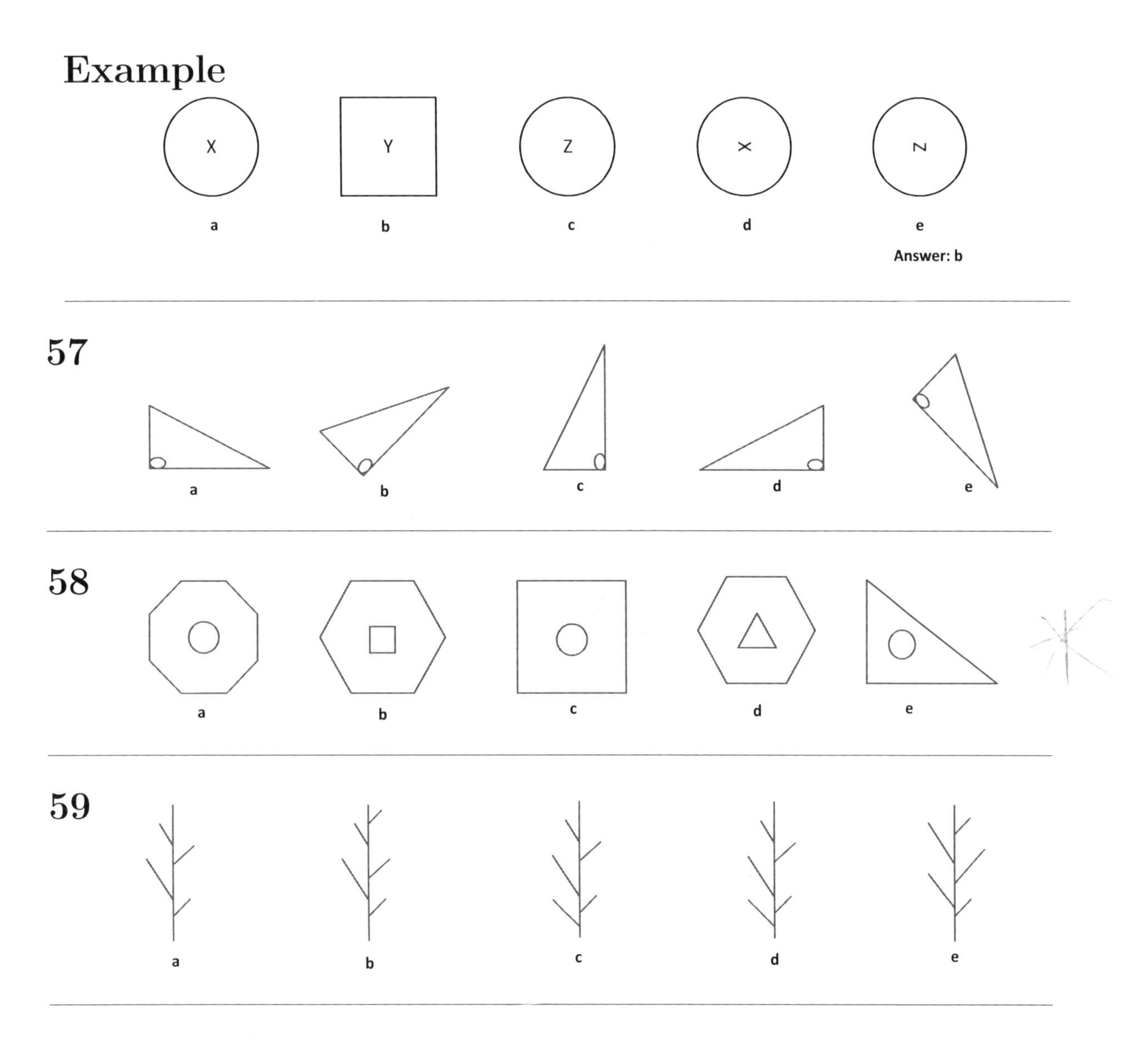

PLEASE CONTINUE TO THE NEXT PAGE

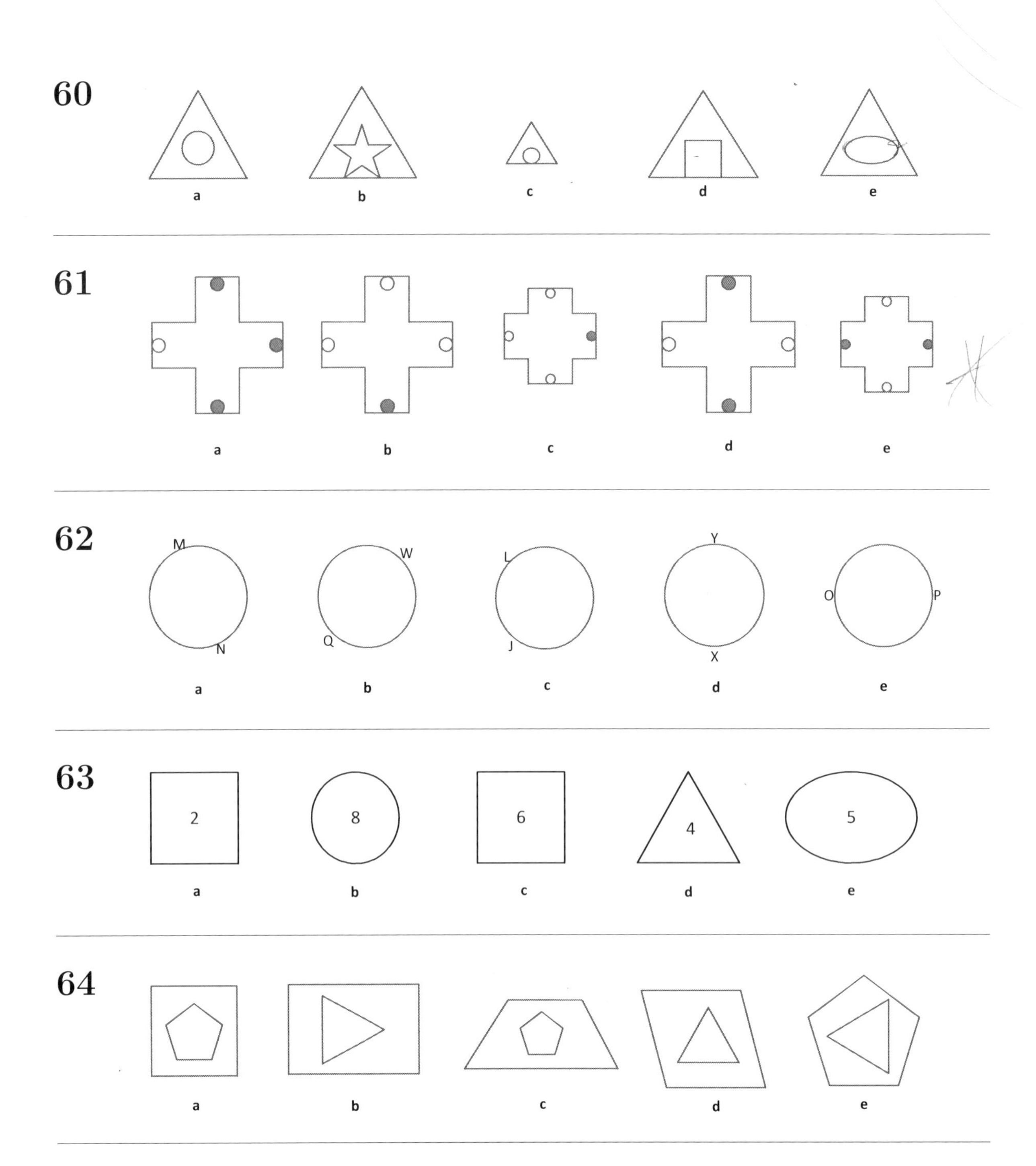

PLEASE CONTINUE TO THE NEXT PAGE

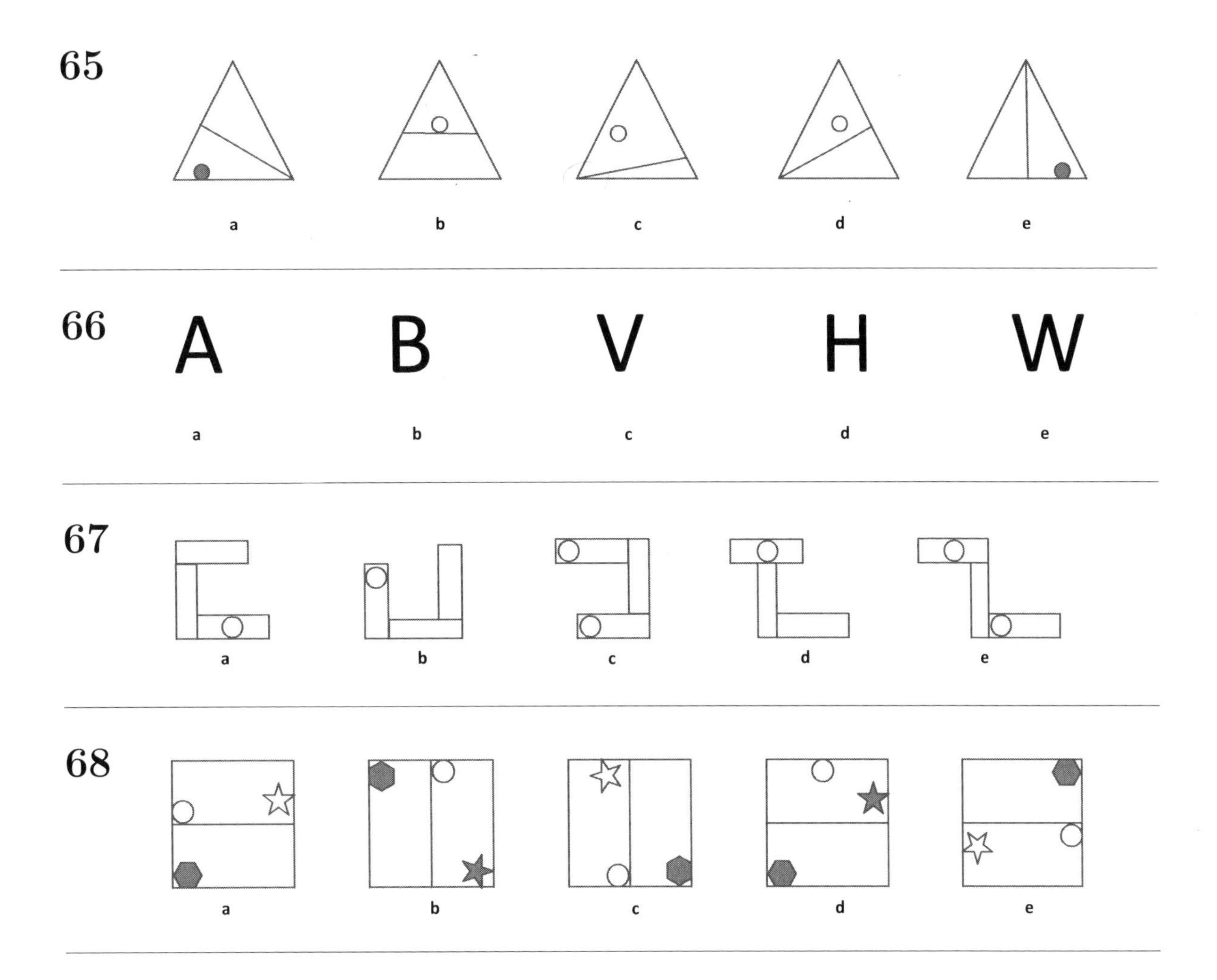

PLEASE CONTINUE TO THE NEXT PAGE

Section 2

There are two shapes on the left of each of the rows below. The second is related to the first in some way. There are five shapes on the right. Find which one of the five shapes related to the third shape in the same way as the two on the left. Mark its letter on the answer sheet.

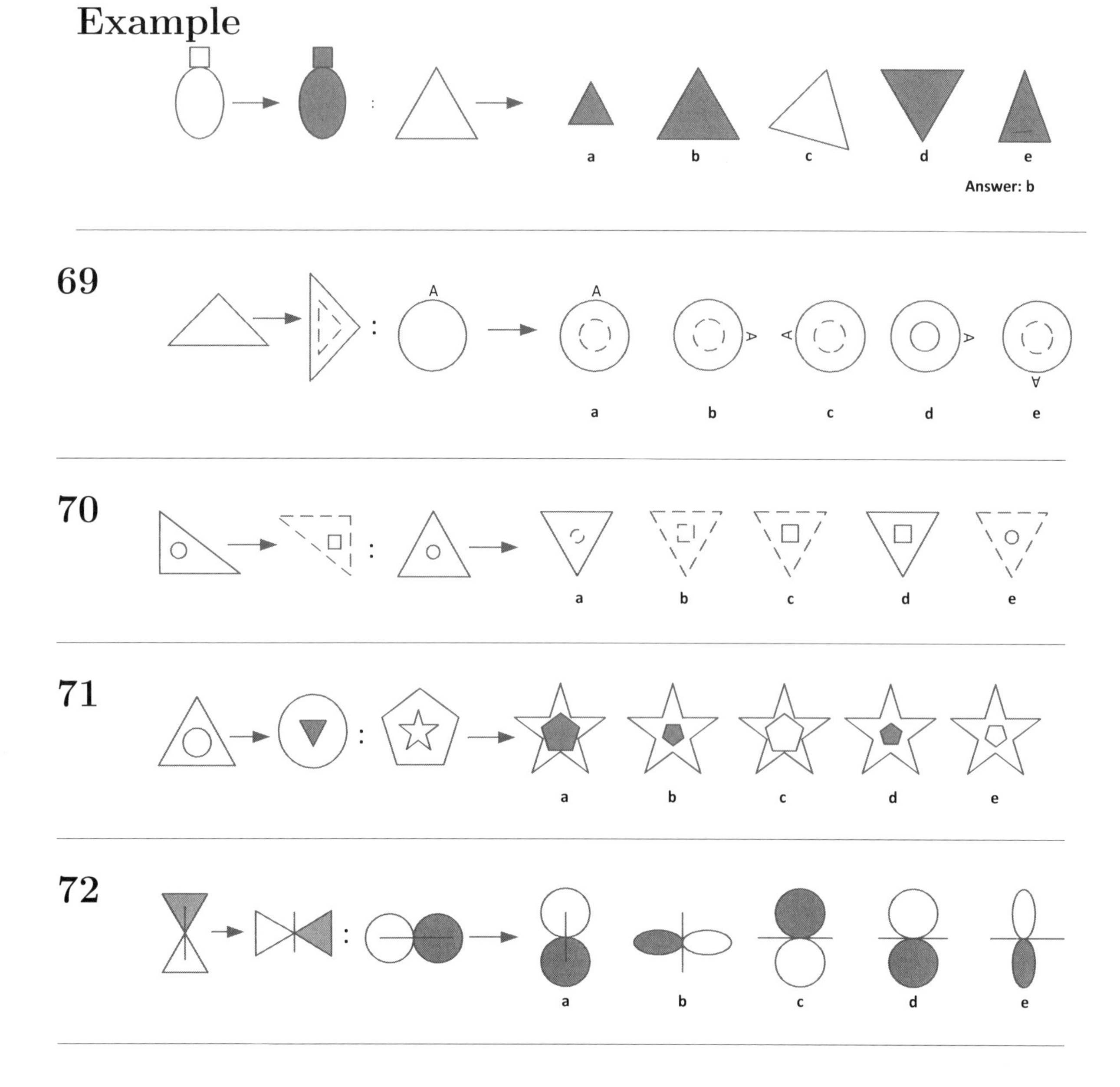

PLEASE CONTINUE TO THE NEXT PAGE

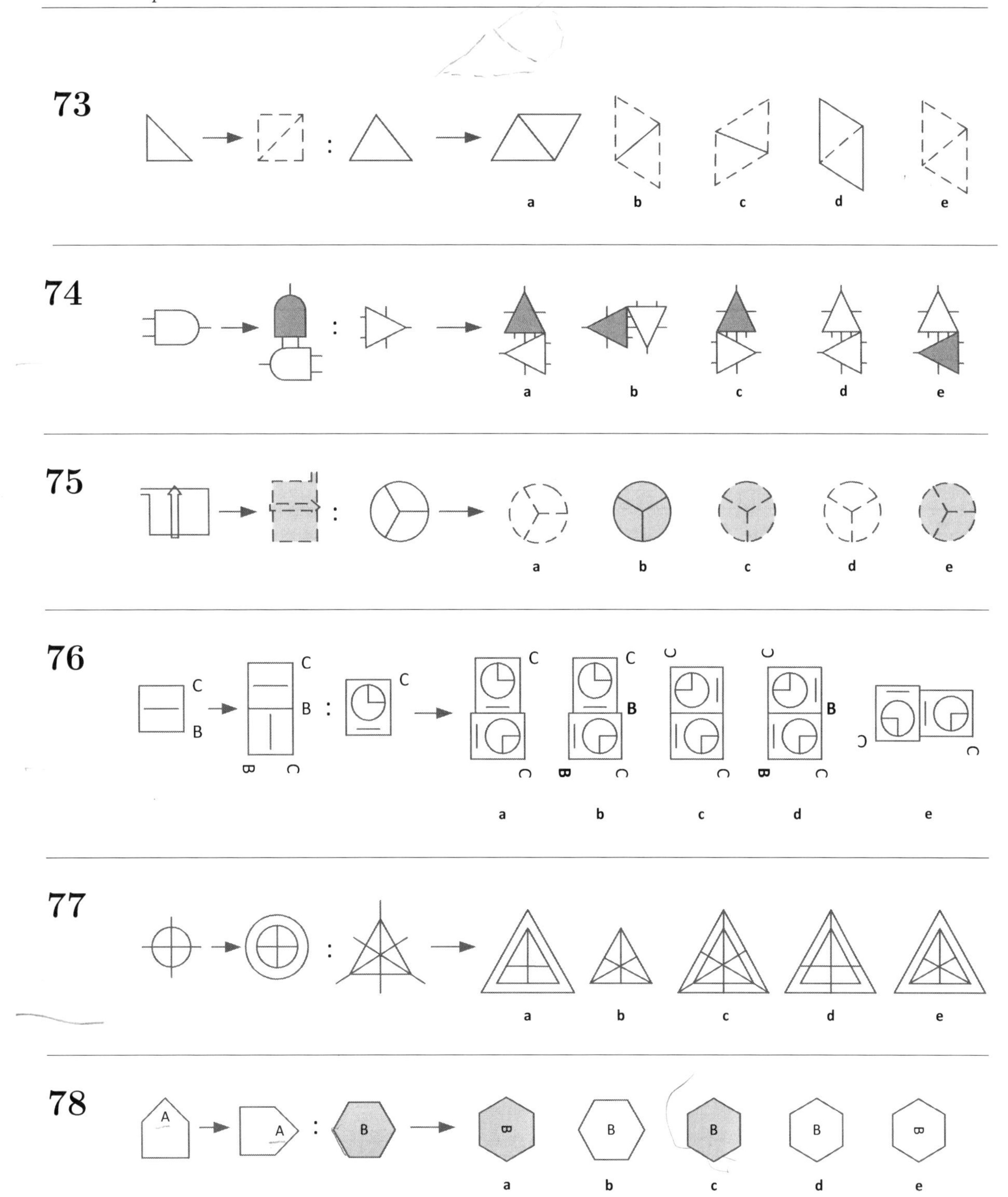

PLEASE CONTINUE TO THE NEXT PAGE

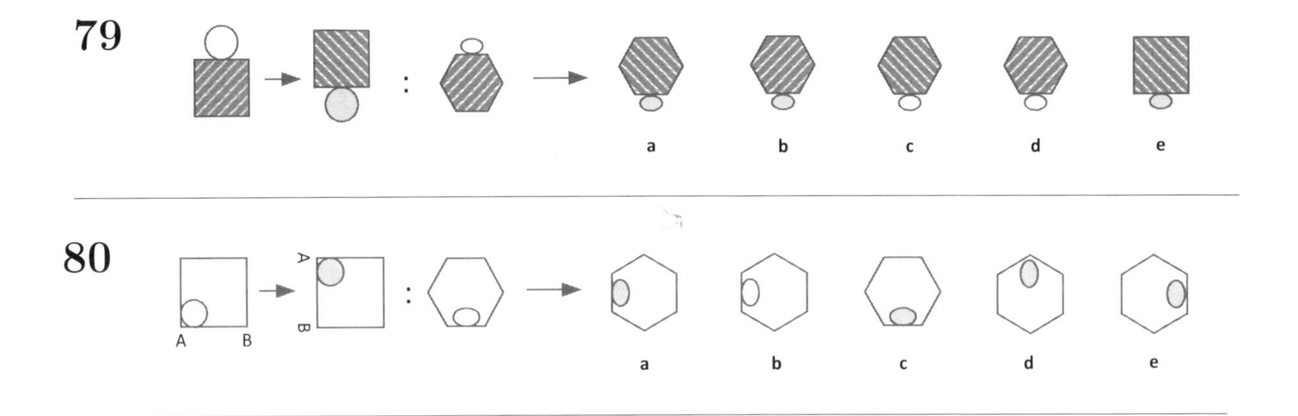

THE END OF NON VERBAL REASONING PART

THE END OF TEST PAPER 11A

11+ Practice Papers

Mixed
Multiple-Choice
Pack Four 11B

Read these instructions before you start:

- There are three sections in this paper.
- You have **50 minutes** in total to complete the whole paper. The time allowed for each section is given at the beginning of that section.
- There are **80 questions** in this paper and each question is worth **one mark**.
- You may work out the answers in rough on a separate sheet of paper.
- Answers should be marked on the answer sheet provided, not on the practice paper.
- Mark your answer in the column marked with the same number as the question by drawing a firm line clearly through the box next to your answer.
- If you make a mistake, rub it out as completely as you can and mark you new answer. You should only mark one answer for each question.
- Work as quickly and carefully as you can.
- If you find a question difficult, do **NOT** spend too much time on it but move on to the next one.
- **Calculators and protractors are not allowed.**

Moon Tuition

making the most of your potential

www.moontuition.co.uk

Part IV

Numerical Reasoning - 25 minutes

Question 1 - 5

black	white	yellow	green	blue
5	4	8	6	7

The table above shows the number of different colour marbles in a box.

The table below is the answer bank for question 1 to 5.
Choose your answers from the answer bank and mark your choice in the answer sheet.

A	B	C	D	E	F	G	H
$\frac{2}{15}$	$\frac{1}{5}$	$\frac{7}{30}$	$\frac{7}{10}$	$\frac{3}{10}$	$\frac{4}{15}$	$\frac{2}{3}$	4:6

1 What is the probability of taking out a blue marble from the box?

---

2 What is the probability of taking out a white marble from the box?

---

3 What is the probability of taking out a green marble from the box?

---

4 What is the proportion of yellow marbles?

---

5 What is the ratio of white marbles to green marbles?

Question 6 - 13

6 What is the digit 5 worth in the number 708.051?

A) 5 B) 0.5 C) 0.05 D) 500 E) 50

---

7 A bus travels 5km in 20 minutes. What is its speed in km/hour?

A) 15 B) 5 C) 10 D) 20 E) 25

---

8 A boat travels 10km in 15 minutes. At the same speed how long will it take the boat to travel 40km?

A) 12 minutes B) 40 minutes C) 30 minutes D) 20 minutes E) 1 hour

---

9 1 pound can be exchanged for 1.5 US dollars. How many US dollars are worth 50 pounds?

A) 30 B) 25 C) 15 D) 75 E) 60

---

10 If 888x666=591408, what is 88.8x66.6?

A) 59140.8 B) 5914.08 C) 591.408 D) 5.91408 E) 59.1408

---

11 What is the lowest common multiple of 40 and 60?

A) 90 B) 60 C) 20 D) 120 E) 240

---

12 What is the value of m if 4m=160?

A) 20 B) 4 C) 40 D) 80 E) 120

---

13 If 4n-5=11, what is the value of n?

A) 2 B) 4 C) 5 D) 8 E) 6

Question 14 - 20

14 If a=2, b=3, c=4, what is a(b+c)?

A) 10 B) 15 C) 16 D) 14 E) 60

---

15 Tom's weekly salary was £60. He just received a 20 % pay rise. How much does he earn per week now ?

A) £48 B) £80 C) £60 D) £62 E) £72

---

16 I am a square number which is bigger than 120 but smaller than 140. What is the sum of my digits ?

A) 3 B) 5 C) 4 D) 6 E) 7

---

17 I am a rectangle with longer side of 12cm and the shorter side of 5cm. What is my area?

A) 34cm B) 15cm C) 17cm D) $60cm^2$ E) 24cm

---

18 A box can hold 6 bottles of water. How many boxes are needed for 122 bottles of water?

A)12 B) 20 C) 19 D) 21 E) 22

---

19 Mr Jones decided to share £40 among his 3 children. He gave Sam $\frac{1}{4}$ of the money and John $\frac{1}{8}$ of the money. How much will the third child get?

A)13 B) 24 C) 10 D) 25 E) 20

---

20 What is the highest common factor of 40 and 80?

A)10 B) 120 C) 240 D) 250 E) 40

Question 21 - 28

21 What is the lowest common multiple of 100 and 500?

A)20 B) 500 C) 200 D) 50 E) 5

---

22 What is the lowest prime common factor of 40 and 60?

A)4 B) 2 C) 5 D) 10 E) 12

---

23 How many quarters are there in 5?

A) 2 B) 5 C) 24 D) 12 E) 20

---

24 What is 75 % of 60?

A) 20 B) 45 C) 15 D) 50 E) 40

---

25 What is difference between 20 % of 50 and $\frac{1}{3}$ of 27?

A) 2 B) 3 C) 1 D) 5 E) 4

---

26 A bag of rice costs £13.50. A bag of flour costs £2.80. How much more expensive is a bag of rice than a bag of flour?

A) £10.70 B) £11.70 C) £10.30 D) £11.30 E) £9.70

---

27 How many factors does 8 have in total?

A) 1 B) 2 C) 3 D) 4 E) 5

---

28 How much bigger is $\frac{5}{6}$ than $\frac{1}{3}$?

A) $\frac{1}{4}$ B) $\frac{1}{2}$ C) $\frac{1}{3}$ D) $\frac{1}{6}$ E) $\frac{2}{3}$

Question 29 - 30

29 I am made from two equal squares placed side by side. Their total area is $50cm^2$. What is the perimeter of one of the squares?

A) 25cm B) 10cm C) 20cm D) 40cm E) 24cm

---

30 Tom travels from his home to his friend Richard's house which is 5km away by a taxi. The taxi driver charges him £2 for the first 3km and 40p for each additional 500 metres. How much does it cost in total?

A) £2.40 B) £2.60 C) £2.80 D) £3 E) £3.60

THE END OF NUMERICAL REASONING PART

Part V
Verbal Reasoning - 15 minutes

Section 1 - Antonyms

Select the word that means the OPPOSITE of the word in bold. Mark your answer on the answer sheet by choosing one of the options A-E.

31

minute

A. large B. tiny C. fine D. small E. fixed

32

rapidly

A. swiftly B. slowly C. ruffled D. weaken E. suddenly

33

triumphant

A. excited B. elated C. unsuccessful D. speedy E. preen

34

slender

A. thick B. slim C. thin D. concerned E. faulty

35

steady

A. discontinuous B. even C. stable D. regular E. smooth

PLEASE CONTINUE TO THE NEXT PAGE

36

agile

A. inactive B. active C. flexible D. running E. continue

37

prior

A. proceed B. after C.high D.weave E. uneven

38

disclose

A. open B. reveal C. hide D. settle E. damp

PLEASE CONTINUE TO THE NEXT PAGE

Section 2 - Comprehension Test

The following extract is taken from the novel, David Copperfield. Read through it and then answer the questions that follow.

'I am David Copperfield, of Blunderstone, in Suffolk. I have been very unhappy since Mother died. I have been slighted, and taught nothing, and thrown upon myself, and put to work not fit for me. It made me run away to you. I was robbed at first setting out, and have walked all the way, and have never slept in a bed since I began the journey.' Here my self-support gave way all at once; and with a movement of my hands, I broke into a passion of crying.

My aunt got up in a great hurry, collared me and took me into the parlour. Her first proceeding there was to unlock a tall press, bring out several bottles, and pour some of the contents of each into my mouth. I think they must have been taken out at random, for I am sure I tasted aniseed water, anchovy sauce, and salad dressing. Then she put me on the sofa, with a shawl under my head, and the handkerchief from her own head under my feet, lest I should sully the cover.

After a time she rang the bell. 'Janet,' said my aunt, when her servant came in. 'Go upstairs, give my compliments to Mr. Dick, and say I wish to speak to him.'

Janet looked a little surprised to see me lying stiffly on the sofa (I was afraid to move lest it should be displeasing to my aunt), but went on her errand.

'Mr. Dick,' said my aunt to him when he had arrived, 'you have heard me mention David Copperfield? Now don't pretend not to have a memory, because you and I know better.'

'David Copperfield?' said Mr. Dick, who did not appear to me to remember much about me. 'David Copperfield? Oh yes, to be sure. David, certainly.'

'Well then,' returned my aunt, 'Now, here you see young David Copperfield, and the question I put to you is, what shall I do with him?'

'What shall you do with him?' said Mr. Dick, feebly, scratching his head. 'Oh! Do with him?'

'Yes,' said my aunt, with a grave look, and her forefinger held up. 'Come! I want some very sound advice.'

'Why, if I was you,' said Mr. Dick, considering, and looking vacantly at me, 'I should -' The contemplation of me seemed to inspire him with a sudden idea, and he added, briskly, 'I should wash him!'

39 Which of the following is TRUE?

A)David Copperfield is a servant.
B)David Copperfield is an old friend of M. Dick.
C)I had a very comfortable bed.
D)I have always been a very happy child.
E)I have never had a job which suits me.

---

40 The word 'Suffolk' in line 1 is a:

A)Participle
B)Preposition
C)Adverb
D)Adjective
E)Proper noun

---

41 The word 'very' in line 1 is a:

A)Participle
B)Preposition
C)Adverb
D)Adjective
E)Proper noun

---

42 The word 'unhappy' in line 1 is a:

A)Participle
B)Preposition
C)Adverb
D)Adjective
E)Proper noun

---

43 Why did David run to his aunt?

A)To pay a visit.
B)To celebrate his aunt's birthday.
C)Because he had been slighted, and taught nothing, and put to work not fit for him.
D)Because his mother asked him to.
E)Because he is an orphan now.

---

44 What happened to David on the journey to his aunt?

A)He had a great time.
B)He was in a train for hours.
C)he was robbed at first setting.
D)He was homesick.
E)He regretted for running away.

---

45 Which of the following occurs in the second paragraph?

A)My aunt put me on the chair.
B)The tall press was left open before my aunt took out several bottles.
C)I was taken to the parlour by my aunt.
D)I wasn't given anything from the bottles.
E)My aunt put a handkerchief on my head.

---

46 The meaning of the word 'collared' in the context 'My aunt got up in a great hurry, collared me and took me into the parlour.' is:

A)shirt
B)coat
C)dress
D)seized by the collar
E)leather

---

47 Why did Janet look a little surprised when she came in?

A)Because David was lying stiffly on the sofa.
B)Because she was reluctant to do as she was told.
C)Because she was not happy to speak to Mr. Dick.
D)Because she didn't like David.
E)Because David's aunt didn't introduce David to her.

---

48 Which of the following statements is NOT TRUE?

A)David's aunt asked Mr. Dick for some advice about how to do with David.
B)Mr. Dick didn't seem to remember David at first.
C)Mr. Dick suggested to wash David.
D)David's aunt didn't get any advice from Mr. Dick.
E)David's aunt believed she had mentioned David to Mr. Dick before.

---

49 The word 'scratching' in line 'Mr. Dick, feebly, scratching his head.' is a:

A)Participle
B)Preposition
C)Adverb
D)Adjective
E)Proper noun

---

50 The meaning of the word 'briskly' in line 'The contemplation of me seemed to inspire him with a sudden idea, and he added, briskly.' is:

A)weakly
B)angrily
C)uncertainly
D)concerned
E)quickly

PLEASE CONTINUE TO THE NEXT PAGE

Section 3 - Missing word

In each of the following sentences, there is a word missing. Please complete each sentence by selecting one word from the options A-E. Mark your answer on the answer sheet.

51 The girl is (   ) good at drawing but also very talented in singing.

A) only B) not only C) merely D) slightly E) purely

---

52 This movie is just (   ) good to be missed.

A) no B) not C) too D) to E) so

---

53 David was lying stiffly (   ) his aunt wouldn't be offended.

A) so that B) such that C) So D) such E) although

---

54 I suddenly felt acute (   ) in my feet while I was running in the fields.

A) spine B) soil C)pains D) sand E) extend

---

55 Feeling very drowsy, I (   ) myself awake for just 5 minutes.

A) can keep B) could keep C)couldn't keep D) couldn't E) be manage

---

56 First of all, you must always avoid (   ) at the gear stick.

A) be looking B) looking C)looked D) make E) ruffle

THE END OF VERBAL REASONING PART

Part VI
Non Verbal Reasoning - 10 minutes

Section 1

Which one of the five shapes completes the sequence? Mark its letter on the answer sheet.

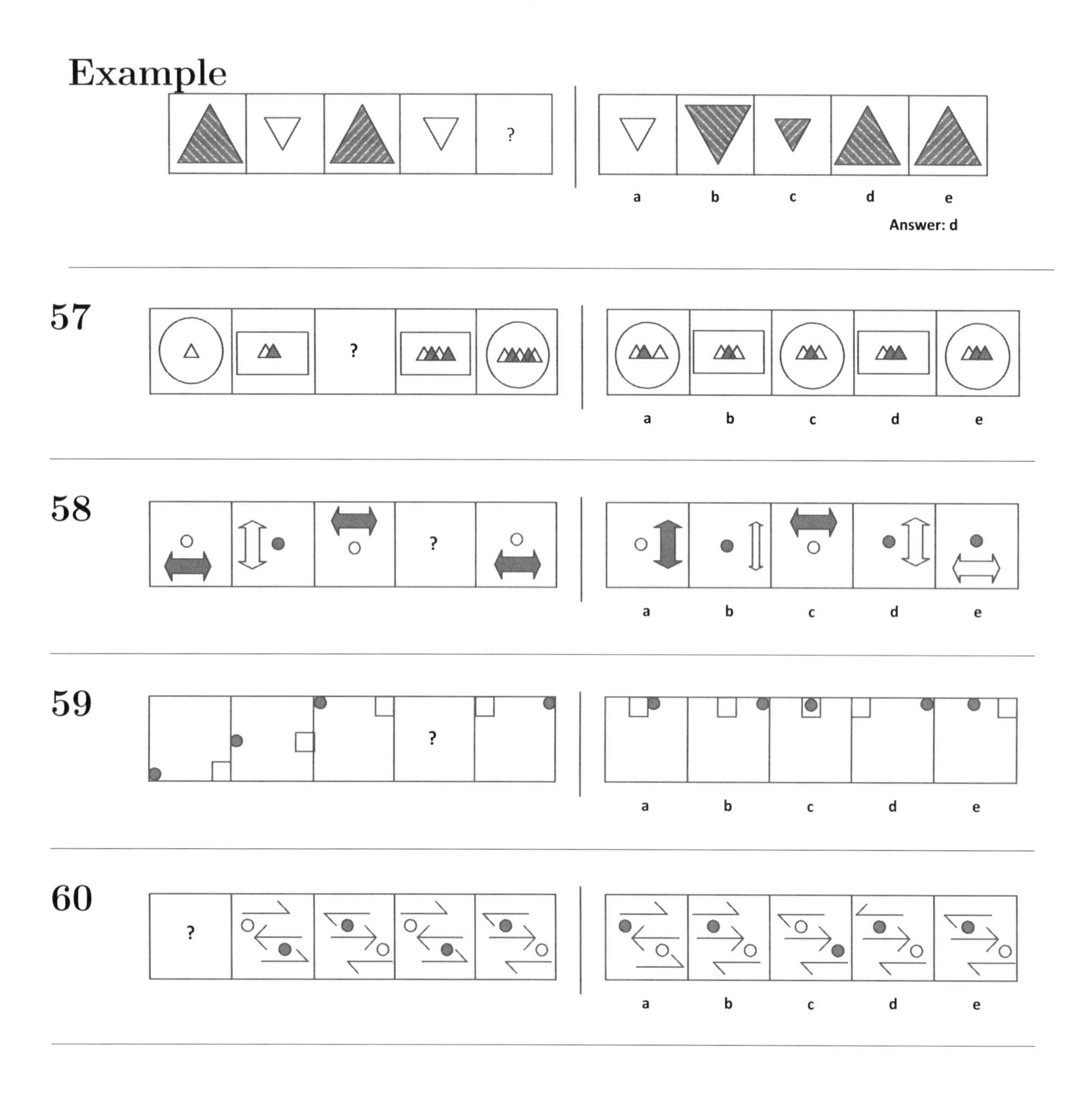

PLEASE CONTINUE TO THE NEXT PAGE

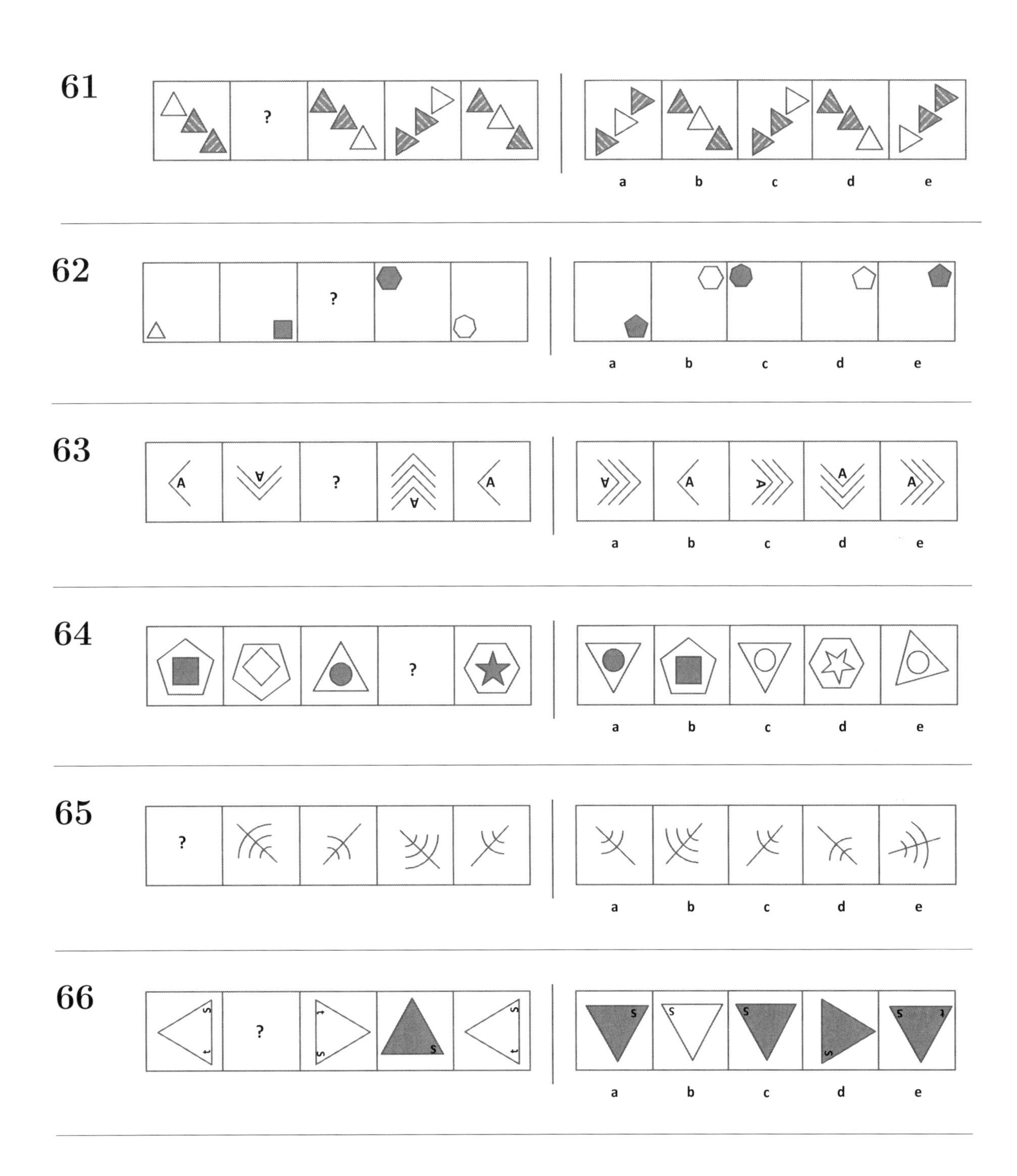

PLEASE CONTINUE TO THE NEXT PAGE

67

68

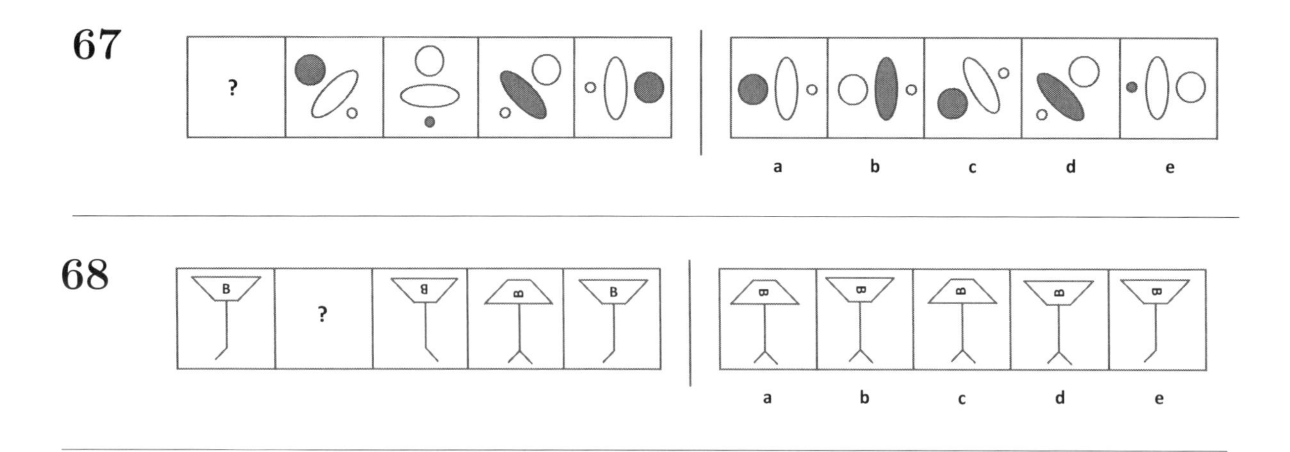

PLEASE CONTINUE TO THE NEXT PAGE

Section 2

There are smaller squares in each large square on the left. One of the small squares has been left empty. Work out which one of the five figures on the right can fill the empty square and mark its letter on the answer sheet.

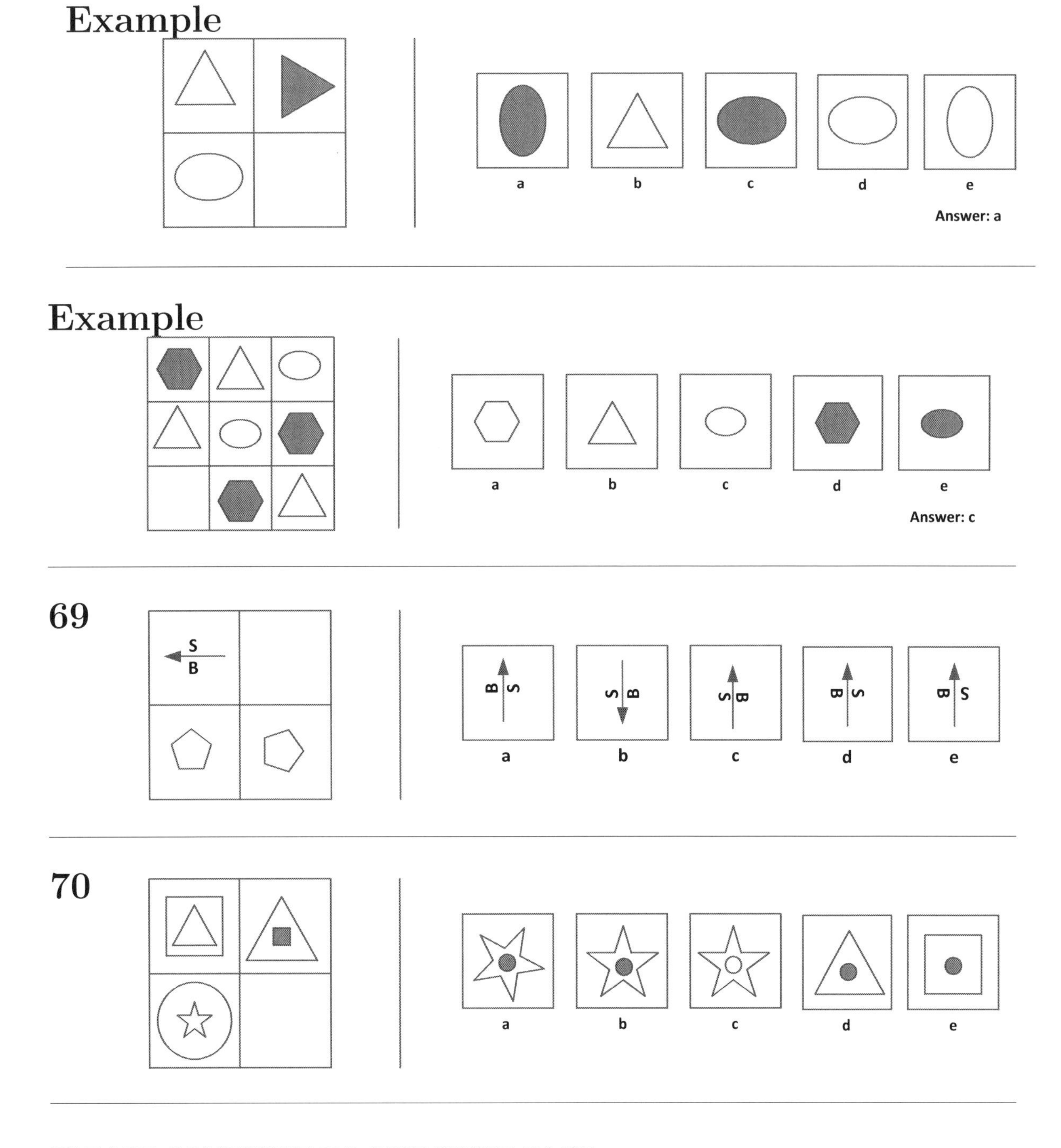

PLEASE CONTINUE TO THE NEXT PAGE

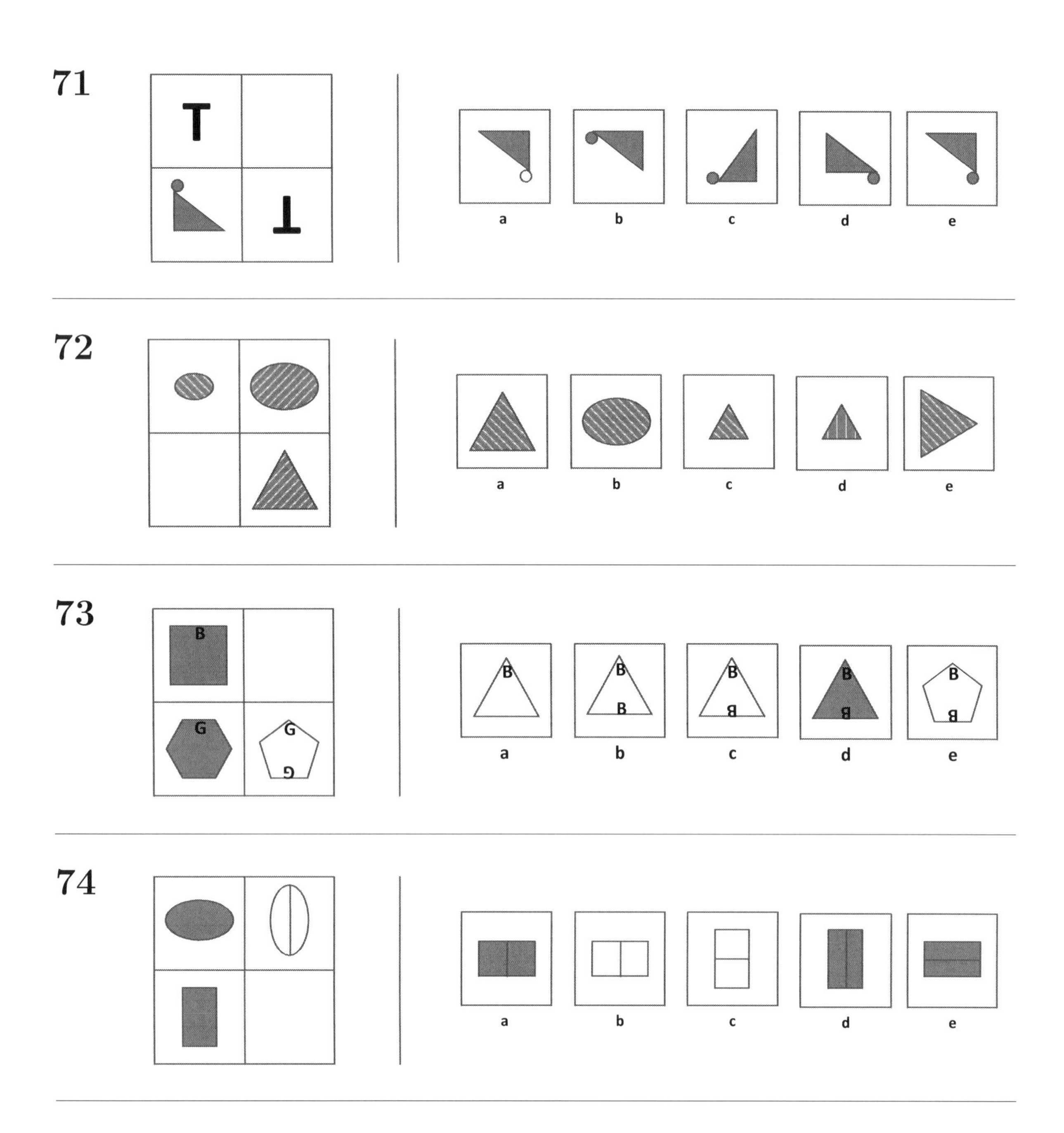

PLEASE CONTINUE TO THE NEXT PAGE

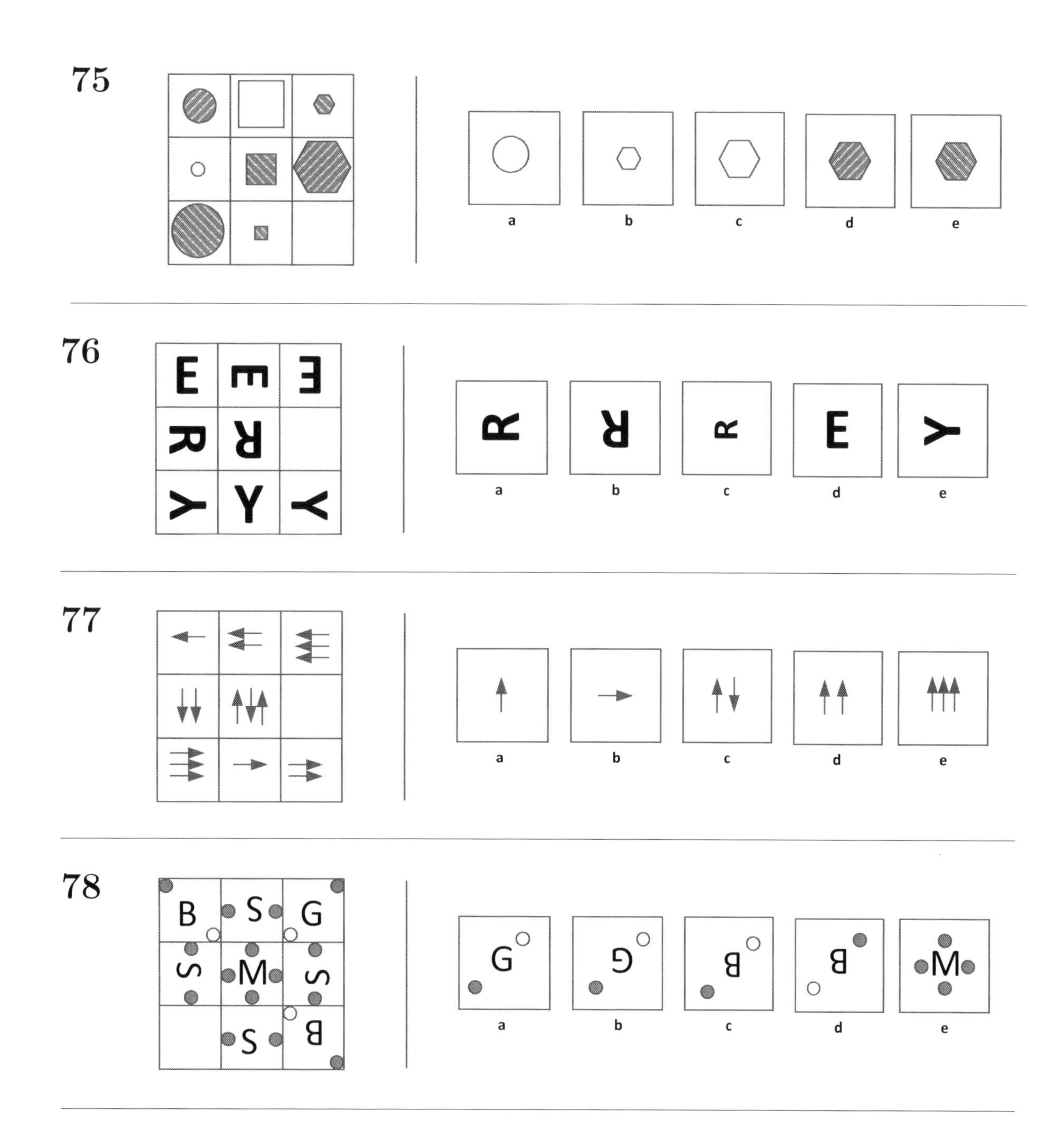

PLEASE CONTINUE TO THE NEXT PAGE

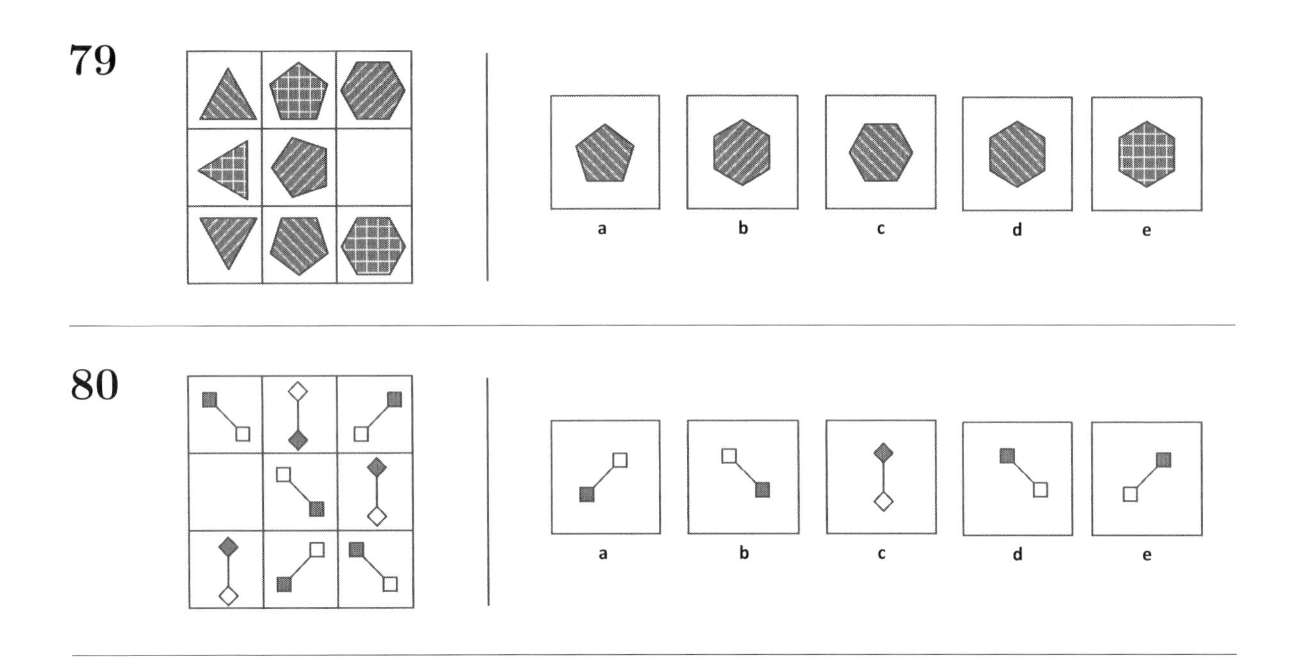

THE END OF NON VERBAL REASONING PART

THE END OF TEST PAPER 11B

11+ Answer Sheets

Multiple-Choice Practice Papers
Pack One

The following answer sheets are included:

- Multiple-Choice Practice Paper Pack One 11A.
- Multiple-Choice Practice Paper Pack One 11B.

Moon Tuition

making the most of your potential

www.moontuition.co.uk

Pack Four 11A - Answer Sheet Page 1

Pupil's Name

School's Name

Date of Test

Please mark like this ⟵]

PUPIL NUMBER

0	0	0	0	0	0
1	1	1	1	1	1
2	2	2	2	2	2
3	3	3	3	3	3
4	4	4	4	4	4
5	5	5	5	5	5
6	6	6	6	6	6
7	7	7	7	7	7
8	8	8	8	8	8
9	9	9	9	9	9

SCHOOL NUMBER

0	0	0	0	0	0	0
1	1	1	1	1	1	1
2	2	2	2	2	2	2
3	3	3	3	3	3	3
4	4	4	4	4	4	4
5	5	5	5	5	5	5
6	6	6	6	6	6	6
7	7	7	7	7	7	7
8	8	8	8	8	8	8
9	9	9	9	9	9	9

DATE OF BIRTH

Day		Month	Year
0	0	January	2000
1	1	February	2001
2	2	March	2002
3	3	April	2003
	4	May	2004
	5	June	2005
	6	July	2006
	7	August	2007
	8	September	2008
	9	October	2009
		November	2010
		December	2011

1 A B C D E F G H

2 A B C D E F G H

3 A B C D E F G H

4 A B C D E F G H

5 A B C D E F G H

6 A B C D E F G H I J K L

7 A B C D E F G H I J K L

8 A B C D E F G H I J K L

9 A B C D E F G H I J K L

10 A B C D E F G H I J K L

11 A B C D E F G H I J K L

12 A B C D E F G H I J K L

13 A B C D E F G H I J

14 A B C D E F G H I J

15 A B C D E F G H I J

16 A B C D E F G H I J

17 A B C D E F G H I J

18 A B C D E F G H I J

19 A B C D E F G H I J

20 A B C D E F G H I J

21 A B C D E F G H I J

22 A B C D E F G H I J

23 A B C D E F G H I J

24 A B C D E F G H I J

25 A B C D E F G H I J

26 A B C D E F G H I J

27 A B C D E

28 A B C D E

29 A B C D E

30 A B C D E

Moon Tuition
making the most of your potential

Pack Four 11A - Answer Sheet Page 2

Pupil's Name

School's Name

Date of Test

DATE OF BIRTH

Day		Month	Year
0	0	January	2000
1	1	February	2001
2	2	March	2002
3	3	April	2003
	4	May	2004
	5	June	2005
	6	July	2006
	7	August	2007
	8	September	2008
	9	October	2009
		November	2010
		December	2011

PUPIL NUMBER

0	0	0	0	0	0
1	1	1	1	1	1
2	2	2	2	2	2
3	3	3	3	3	3
4	4	4	4	4	4
5	5	5	5	5	5
6	6	6	6	6	6
7	7	7	7	7	7
8	8	8	8	8	8
9	9	9	9	9	9

SCHOOL NUMBER

0	0	0	0	0	0	0
1	1	1	1	1	1	1
2	2	2	2	2	2	2
3	3	3	3	3	3	3
4	4	4	4	4	4	4
5	5	5	5	5	5	5
6	6	6	6	6	6	6
7	7	7	7	7	7	7
8	8	8	8	8	8	8
9	9	9	9	9	9	9

Please mark like this ⟵

31 A B C D E
32 A B C D E
33 A B C D E
34 A B C D E
35 A B C D E
36 A B C D E
37 A B C D E
38 A B C D E
39 A B C D E
40 A B C D E
41 A B C D E
42 A B C D E
43 A B C D E
44 A B C D E
45 A B C D E
46 A B C D E
47 A B C D E
48 A B C D E
49 A B C D E
50 A B C D E
51 A B C D E
52 A B C D E
53 A B C D E
54 A B C D E
55 A B C D E
56 A B C D E
57 A B C D E
58 A B C D E
59 A B C D E
60 A B C D E
61 A B C D E
62 A B C D E
63 A B C D E
64 A B C D E
65 A B C D E
66 A B C D E
67 A B C D E
68 A B C D E
69 A B C D E
70 A B C D E
71 A B C D E
72 A B C D E
73 A B C D E
74 A B C D E
75 A B C D E
76 A B C D E
77 A B C D E
78 A B C D E
79 A B C D E
80 A B C D E

Pack Four 11B - Answer Sheet Page 1

Pupil's Name	Date of Test

School's Name

Please mark like this

PUPIL NUMBER

0	0	0	0	0	0
1	1	1	1	1	1
2	2	2	2	2	2
3	3	3	3	3	3
4	4	4	4	4	4
5	5	5	5	5	5
6	6	6	6	6	6
7	7	7	7	7	7
8	8	8	8	8	8
9	9	9	9	9	9

SCHOOL NUMBER

0	0	0	0	0	0	0
1	1	1	1	1	1	1
2	2	2	2	2	2	2
3	3	3	3	3	3	3
4	4	4	4	4	4	4
5	5	5	5	5	5	5
6	6	6	6	6	6	6
7	7	7	7	7	7	7
8	8	8	8	8	8	8
9	9	9	9	9	9	9

DATE OF BIRTH

Day		Month	Year
0	0	January	2000
1	1	February	2001
2	2	March	2002
3	3	April	2003
	4	May	2004
	5	June	2005
	6	July	2006
	7	August	2007
	8	September	2008
	9	October	2009
		November	2010
		December	2011

1	2	3	4	5
A	A	A	A	A
B	B	B	B	B
C	C	C	C	C
D	D	D	D	D
E	E	E	E	E
F	F	F	F	F
G	G	G	G	G
H	H	H	H	H

6	7	8	9	10	11	12	13	14
A	A	A	A	A	A	A	A	A
B	B	B	B	B	B	B	B	B
C	C	C	C	C	C	C	C	C
D	D	D	D	D	D	D	D	D
E	E	E	E	E	E	E	E	E

15	16	17	18	19	20	21	22	23
A	A	A	A	A	A	A	A	A
B	B	B	B	B	B	B	B	B
C	C	C	C	C	C	C	C	C
D	D	D	D	D	D	D	D	D
E	E	E	E	E	E	E	E	E

24	25	26	27	28	29	30
A	A	A	A	A	A	A
B	B	B	B	B	B	B
C	C	C	C	C	C	C
D	D	D	D	D	D	D
E	E	E	E	E	E	E

Moon Tuition

making the most of your potential

Pack Four 11B - Answer Sheet Page 2

Pupil's Name

School's Name

Date of Test

Please mark like this ⟵

PUPIL NUMBER					
0	0	0	0	0	0
1	1	1	1	1	1
2	2	2	2	2	2
3	3	3	3	3	3
4	4	4	4	4	4
5	5	5	5	5	5
6	6	6	6	6	6
7	7	7	7	7	7
8	8	8	8	8	8
9	9	9	9	9	9

SCHOOL NUMBER						
0	0	0	0	0	0	0
1	1	1	1	1	1	1
2	2	2	2	2	2	2
3	3	3	3	3	3	3
4	4	4	4	4	4	4
5	5	5	5	5	5	5
6	6	6	6	6	6	6
7	7	7	7	7	7	7
8	8	8	8	8	8	8
9	9	9	9	9	9	9

DATE OF BIRTH			
Day		Month	Year
0	0	January	2000
1	1	February	2001
2	2	March	2002
3	3	April	2003
	4	May	2004
	5	June	2005
	6	July	2006
	7	August	2007
	8	September	2008
	9	October	2009
		November	2010
		December	2011

31 A B C D E
32 A B C D E
33 A B C D E
34 A B C D E
35 A B C D E
36 A B C D E
37 A B C D E
38 A B C D E
39 A B C D E
40 A B C D E
41 A B C D E
42 A B C D E
43 A B C D E
44 A B C D E
45 A B C D E
46 A B C D E
47 A B C D E
48 A B C D E
49 A B C D E
50 A B C D E
51 A B C D E
52 A B C D E
53 A B C D E
54 A B C D E
55 A B C D E
56 A B C D E
57 A B C D E
58 A B C D E
59 A B C D E
60 A B C D E
61 A B C D E
62 A B C D E
63 A B C D E
64 A B C D E
65 A B C D E
66 A B C D E
67 A B C D E
68 A B C D E
69 A B C D E
70 A B C D E
71 A B C D E
72 A B C D E
73 A B C D E
74 A B C D E
75 A B C D E
76 A B C D E
77 A B C D E
78 A B C D E
79 A B C D E
80 A B C D E

11+ Answer Key

Multiple-Choice Practice Papers
Pack Four

Read these instructions before you start marking:

- Only the answers given are allowed.
- One mark should be given for each correct answer.
- Do not deduct marks for the wrong answers.

making the most of your potential

www.moontuition.co.uk

Practice Paper Answers

Practice Paper 11A				Practice Paper 11A			
1.	B	26.	J	51.	D	76.	A
2.	A	27.	C	52.	A	77.	E
3.	D	28.	B	53.	E	78.	C
4.	C	29.	D	54.	B	79.	A
5.	F	30.	E	55.	A	80.	A
6.	C	31.	A	56.	D		
7.	D	32.	C	57.	D		
8.	A	33.	B	58.	E		
9.	H	34.	D	59.	A		
10.	I	35.	A	60.	C		
11.	J	36.	C	61.	E		
12.	K	37.	B	62.	C		
13.	B	38.	A	63.	E		
14.	C	39.	B	64.	E		
15.	F	40.	D	65.	B		
16.	H	41.	C	66.	B		
17.	A	42.	E	67.	D		
18.	J	43.	B	68.	D		
19.	E	44.	E	69.	B		
20.	E	45.	B	70.	C		
21.	B	46.	A	71.	B		
22.	G	47.	C	72.	D		
23.	H	48.	E	73.	E		
24.	I	49.	B	74.	A		
25.	F	50.	A	75.	C		

Practice Paper Answers

Practice Paper 11B				Practice Paper 11B			
1.	C	26.	A	51.	B	76.	A
2.	A	27.	D	52.	C	77.	A
3.	B	28.	B	53.	A	78.	B
4.	F	29.	C	54.	C	79.	D
5.	H	30.	E	55.	C	80.	E
6.	C	31.	A	56.	B		
7.	A	32.	B	57.	C		
8.	E	33.	C	58.	D		
9.	D	34.	A	59.	C		
10.	B	35.	A	60.	E		
11.	D	36.	A	61.	A		
12.	C	37.	B	62.	D		
13.	B	38.	C	63.	E		
14.	D	39.	E	64.	C		
15.	E	40.	E	65.	C		
16.	C	41.	C	66.	C		
17.	D	42.	D	67.	B		
18.	D	43.	C	68.	A		
19.	D	44.	C	69.	D		
20.	E	45.	C	70.	B		
21.	B	46.	D	71.	E		
22.	B	47.	A	72.	C		
23.	E	48.	D	73.	C		
24.	B	49.	A	74.	B		
25.	C	50.	E	75.	C		

Printed in Great Britain
by Amazon